The World Ends On The 9th

The World Ends On The 9th

Calculating When WWIII Will Begin

Dylan Clearfield

G. Stempien Publishing Company

THE WORLD ENDS ON THE 9TH:
Calculating When WWIII Will Begin
(Algorithm Based on Previous Doomsday Events)
BY
DYLAN CLEARFIELD

Everything is reducible to a common factor

G. Stempien Publishing Company
Copyright © 2023 by Prism Thomas
ISBN 978-0-930472-67-2
All rights reserved
(Second edition)

CONTENTS

Introduction

Do not take the title literally. It may only be a close approximation. But there is genuine truth in the prediction.

When will nuclear war begin? It's a question on most people's minds. Especially today. Nuclear war seems inevitable. Since the day the atomic bomb was created all of mankind was suddenly on the verge of total annihilation. And since that day of discovery there have been many instances when nuclear war had almost been unleashed.

Can the precise day when this war begins be predicted? It will be attempted here. It may be possible to determine the most likely date when nuclear war will begin by using existing data to predict this cataclysm.

The findings won't be based on religious prophecies. Nor will they be based on psychic conjurations or numerology. The predicted date for the start of nuclear destruction will be based on past events aligned with algorithmic calculations - probability factors. Is this really possible? According to the data in this book - it is!

This idea of prognosticating the start of nuclear war was not a contrived one for this book. It wasn't even an idea. The proposal came about during the writing of this book on a different topic: the history of **near** nuclear holocausts. The dates of these events were **NOT PRE-SELECTED**; they were originally selected by event type only. But then the unusual aspect of the dates was noticed and shocked all of us. They were recurring days of the month!! The same days of the month again and again.

Originally, the events of past **near** nuclear devastation were being compiled and evaluated only to examine their historical background. But as the facts mounted, one inescapable conclusion made itself clear

- these events occurred at various specific times in the past. Almost as if planned or under some direction. Orchestrated near nuclear catastrophes!!

This book will recount this group of **near** nuclear catastrophes which are seemingly repetitious. What might account for this will be examined at the end of the book. But for as yet unknown reasons, nuclear war seems more likely to occur on days numbered 9, 23, 25 and 27 than any other numbered day of the month. This will be verified by the data to follow.

It is important to note that ONLY events that qualified as creating potential worldwide destruction through nuclear exchange - or were mysteriously avoided - were included in this collection of near cataclysms. Simple hazardous mishaps which were immediately corrected were excluded from the list as were some natural disasters that would not have in themselves triggered a world war - including Chernobyl and Fukushima (Ukraine is still questionable). The list of near nuclear disasters will be examined in chronological order.

As you read through this list of near catastrophes and note the dates on which they occurred you might wish to ask yourself - "Is something controlling these events for some purpose?" Both causing and preventing them? And what of the next time?

It should be pointed out that all of the events to be described in this book were experienced by the author. These are not just vague, newspaper articles or television reports; they are events that have been confronted as they occurred.

Some people claim that portions of civilization will survive a nuclear war. Shelters have been prepared underground to maintain "high society" and politicians.. While this may be true, it is only for the very very very few; and only the wealthy and powerful. The rest of those who survive will return to life as in Neandertal (German spelling) days. There will be no electricity, or grocery stores, or gasoline stations, or....anything.

Outside of Culpeper, Virginia a beautiful mountainside was destroyed in which to insert a bunker for the purpose of housing civilization's entire video and musical history. It was quite a sight watching this subterranean edifice being built. While much of this societal history may possibly survive nuclear war, the question to ask is: who will be left to watch and enjoy it again?

POWER FAILURES & OTHER TRIGGERS

Massive Power failures can be terrifying. When an entire city, state, or region of a country loses all power, this in itself is a major disaster. It's even worse when a major military base is affected.

When the power suddenly fails on a military base it isn't necessarily thrown into darkness which isn't necessarily a good thing; the emergency backup generators take over. And when this happens, any number of accidents on the "line" can occur. One of these accidents might present itself as a false nuclear war warning to the system. Fortunately, most of the people in our military are highly trained and capable of dealing with unusual situations; but not all of them.

When major metropolitan areas are struck by a power failure there are sensors in place that could register this as a nuclear attack. These monitors gauge unusual changes in lighting and if a highly unusual lighting event occurs - like a nuclear detonation - these monitors are likely to send a nuclear attack warning to whatever command system to whichever they are attached.

Such a rare lighting event can also be triggered by unexpected solar output. That is why at least 2 of the near nuclear disaster events to be examined in this book are coronal mass ejections or similar solar eruptions.

Excessively loud noises can have the same effect on monitors, registering the sounds as nuclear explosions. Ironically, a minor nuclear detonation - so called dirty bomb - could be the cause of an alert which precipitates a full scale response from a full force atomic weapon.

There are many factors that can accidentally trigger nuclear war. What follows are the most serious events that have occurred to date (early 2023) in chronological order.

GOLDSBORO DISASTER

Date of occurrence: **January 23, 1961**

This event was inaugurated just before midnight on January **23, 1961**. It happened when the B-52 stratofortress piloted by Walter Scott Tulloch, flying out of Seymour Air Base in North Carolina, made preparations to refuel. This of course would be done while still airborne.

After the proper maneuvers, the stratofortress was connected to the awaiting aerial tanker in what was expected to be a very routine process. But this activity was going to become very different from routine, very quickly, something which made many people very nervous because this craft was armed with 2 nuclear bombs of 3 to 4 megaton explosive power.

Shortly after refueling had commenced, a fuel leak was discovered in the stratofortress's right wing. A bad fuel leak! The plane detached from the tanker and was ordered by ground control to maintain a steady altitude of 30,000 feet until all of the fuel had either been consumed or been drained from the craft.

But the leak was much worse than originally thought. It was massive! In just 3 minutes, 37,000 pounds of fuel had poured out of the craft, making it too unwieldy to fly. So, the pilot was ordered to land as soon as possible.

Descent was immediately begun. When 10,000 feet was reached, the B-52 began to wildly wobble and buck. Control was completely lost. So much so that the pilot ordered the crew to abandon the craft which had begun to be torn apart by the severity of the vibrations.

Five men in total either bailed out in the traditional way or used an ejector seat to escape the doomed B-52. One of the men who escaped by parachute was killed in the descent and no additional information was supplied concerning him.

Two brave men chose to remain with the craft and fight for control of it during its plunge through the sky. Unfortunately, they were both killed in the subsequent crash which scattered debris over a 2 mile mile radius. Among the debris were 2 nuclear bombs of 3 to 4 megaton strength. Both of them were jarred loose during the gyrating dive to the ground with one of the bombs becoming entangled in a parachute and ultimately was lodged in a tree branch.

The second nuclear bomb flipped clear of the B-52 and plunged straight into the muddy, mid winter farmland, drilling downward at a speed of roughly 700 mph. This bomb was buried in the soil to a depth of 20 feet and was finally located by its projecting tail fin.

Had either of these bombs been armed during their uncontrolled descent? According to Lt. Jack Revelle - bomb disposal expert - the bomb that had become lodged in the tree had completed all of its arming functions during the fall - except one. The safety switch which prevented the bomb from fully arming was still in the **ON** position which kept the device from detonating.

That wasn't the case for the other bomb - the one that drilled its way into the ground. It too went through all of the arming procedures during the downward plunge but the safety switch had been thrown from the ON to the OFF position which meant that this device was primed to explode. The bomb now live - primed to detonate.

It didn't. Why not? The armed and primed nuclear bomb did not detonate because during the fall to earth the wire between the detonator and the explosive charge that would have set off the explosion was broken off. The bomb was ready to explode but the "spark" to set it off could not be delivered.

If the bomb had detonated, it would have triggered the other one to explode as well and the result would be the detonation of 2 nuclear bombs of a 3 to 4 megaton force each in the Goldsboro region of North Carolina. The destruction would've been massive and would have alerted the Soviet Union which most likely would have assumed that the explosion was a U.S. ICBM attack on Cuba. Remember, this was 1961 and tracking devices were not nearly as sophisticated as they are today. Also, this was at a time when the Cold War was at its most sensitive and each side was poised for battle. The Cuban Missile Crisis was still a short distance in the future in 1962.

A counterattack would most probably have been launched against America by the Soviet Union if the nuclear bombs that dropped over Goldsboro had detonated. But due to a bit of good fortune on this day of January 23, 1961 this did not occur.

But was this simply good fortune or something more? Was there some type of "intelligence" behind the actions that prevented the nuclear bombs from exploding? And would this be the same "intelligence" that might have also caused the event to happen in the first place?

This might seem a strange question at this early stage of this book. But it will make more sense to the reader as the cases of near nuclear mishap and escape add up.

FEBRUARY 4, 1962

Date of planetary alignment: **February 4, 1962**

The near nuclear disaster events in this book are highlighted and arranged according to date. The purpose is to factually reveal the shared dates that these rare and potentially world changing events share. Considering the limited number of such potential worldwide disasters that exist, it is highly suspicious as to why so many of them occurred on the same day of the month as will be demonstrated in these pages.

With this in mind, it is important to note that among the days of the month that will be spotlighted as being in this category of repetitious days of occurrence, the 4th day of the month stands alone among the group as not being among these repetitious days. On only one occasion was a near worldwide catastrophe avoided on the 4th day of the month - February 4, 1962.

It is included here because of its importance, its timing in world events, and because of its planet wide observance. Observance? And, because of its relation to a situation called apocatastasis, which will be described shortly.

Due to a predicted alignment of all of the planets in our solar system on this one day, the end of the world was prophesied for February 4th in unison by all the most respected prophets, astrologers, tarot card readers, swamis, tea leaf readers, psychics and other soothsayers. Foremost among them was the well known psychic, Lydia Emma Pinckert - much better known as Jeane Dixon.

At the time that this prediction was announced it was a major news item around the world. Because it involved a genuine cosmic event - the alignment of all of the planets - the end of the world was accepted as a possibility because no one knew what effect such an astronomical occurrence could have. The world was genuinely fearful on this day which I know from direct experience.

Although the world didn't end on February 4, 1962, this prediction seemed as if only slightly amiss in time. The much more serious Cuban Missile Crisis lurked only 8 months in the future. And the Goldsboro incident was not that long ago in the past.

Could it be that the earth was passing through a point in space at this time which was favorable for worldwide cataclysmic events? There is a term for this and it is **apocatastasis**. A simple definition of this word is that the elements of future events are present in the locality of outer space through which our solar system passes and this area can affect historical events. During one point in time, our solar system in its trek through the universe had once before passed through this location with dire results and is influenced by these past events during its current passage through this same point in space - **apocatastasis**. Apocatastasis might also be viewed as a form of recycled time.

THE CUBAN MISSILE CRISIS

Dates of Occurrence: **October 23 - 27, 1962**

The Cuban Missile Crisis was an ongoing incident, spanning many weeks, but reached its critical crisis point between October 23 and October 27, 1962. As such, the dates of the 23rd, 24th, 25th, 26, and 27th coincide with a sequential list of independent dates from other near worldwide holocausts. Note that it also includes the **23rd day** of the month!! A mere coincidence?

For 5 straight days during the Cuban Missile Crisis the world teetered on the edge of total destruction. At any moment either side could've launched a nuclear attack. And that would have been the end of the world. As the brilliant quantum physicist Hugh Everett III proved through his detailed calculations, any type of nuclear exchange would ultimately unleash Mutual Assured Destruction.

From October 23 through October 27, 1962 there prevailed constant peril when the slightest mistake, accident or even omission could provoke World War III. If you didn't have a bomb shelter - which few of us did back then - your only place to hide was the public fallout shelters which were usually the sub-basements of department stores or town administration buildings. And everyone knew that this would not provide protection, just extend one's time of suffering from the fallout to come.

But why is the crisis point singled out to this span of dates, starting on the 23rd? Because that is when the U.S. blockade of Cuba began.

The Cuban Missile Crisis was in reality a week's long incident which centered around the Soviet Union secretly installing operational ICBM bases in Cuba. This meant that nuclear bombs could reach the interior of the United States within minutes.

Ironically, should an attack be waged on the US from Cuba, Cuba would have been poisoned by as much radioactive fallout as America from the firing of their own weapons. But, of course, Cuba would be pulverized by an American counterstrike so would not have to worry about radioactive fallout from its own weapons.

Not only had the Soviet Union constructed bases in Cuba but it was actively creating more. It's when President John Kennedy decided to physically put a halt to this supply chain that the climax of the crisis was reached. On the night of October 22nd the president ordered the U.S. Navy to form a blockade of the Island of Cuba, preventing any Soviet ships from passing through to the island if bearing any type of munitions. His television broadcast on the night of October 22 was a historic event. The blockade took effect the next day - **October 23rd.**

To counter the blockade, the Soviets dispatched their Atlantic based submarine fleet to harass American ships from beneath the waves.

At the same time, the United States was dispatching aerial reconnaissance over Cuba, keeping close surveillance on any developments. And this led to one of the most serious encounters. Near the noon hour on October 27 a U2-F class spy plane piloted by Major Rudolf Answerson was shot down from Cuba by an SA-2 surface-to-air missile. The pilot was killed in the subsequent crash.

With tensions as high as they were, it's remarkable that this act did not set off world war III. And this was only 1 of 3 perilous actions that took place on the 27th, a day which came to be known as "Black Saturday."

In order to deflect the Soviet submarines, American destroyers lobbed "signaling" depth charges (practice, unarmed depth charges) above known locations of the enemy sub-surface craft. One of the Soviet submarines was a B-29 type that was armed with nuclear torpedoes and the officers on board were authorized to use them if they believed their ship was in danger. Captain Valentin Grigorievitch was anxious to start a "shooting war" for real, but according to policy had to get the agreement of 2 other officers of the fleet. He had one other in agreement but could not convince the third man, Vasily Arkipov.

On the same day a third event occurred which could easily have triggered war. An American U-2 spy plane made an unauthorized 90 minute flight over the Soviet Union which attracted the attention of MiG jet fighters. In response, the U.S. dispatched a squad of nuclear armed F-102's to intercept the MiGs.

Luckily, the U-2 spy plane exitted Soviet air space before the fighters could make contact with one another and the catastrophe was avoided. Thus came a close to "Black Saturday" with the world somehow still being intact.

And, suddenly - the *entire crisis* was over! Just like that. On October 28th the world awoke to learn of an agreement made between the United States and the Soviet Union that evaded nuclear war. The Soviet Union agreed to remove its nuclear bases from Cuba if the United States removed its Jupiter rocket bases from Turkey and agreed to never invade Cuba.

The span of dates - the **23rd** through the **27th** - must be included in the list of potential world ending events because each day the planet was on the brink of nuclear annihilation at any moment. Again, this event includes these days: **23rd, 24th, 25th, 26th and 27th**. Of these numbered days, the 23rd, 25th and 26th keep recurring.

A RUSSIAN BEAR IN DULUTH?

Date of Occurrence: **October 25, 1962**

This event occurred during the midst of the Cuban Missile Crisis, but in another part of the United States and is not directly related to those specific happenings. Thus, this rates as an individual occurrence and becomes the second near nuclear holocaust to take place on the **25th** day of a month. It was caused by a bear and, but for the quick thinking of one man, might have precipitated nuclear war.

This affair also occurred near midnight (like the Goldsboro event) on **October 25**, 1962. The location was the Volk Air Force Base in Duluth, Minnesota. The fence-enclosed perimeter of this nuclear facility was guarded by armed service personnel 24 hours a day. It was while one of the guards was in close patrol of the security fence that he caught sight of a shadowy figure brushing up against the outside of the fence. The figure then flung itself against the fence, setting off the alarm and causing the guard to fire at him. The intruder - which later turned out to be a huge Minnesota black bear - fled into the night. The alarm system that this curious bear triggered was electronically connected to the alarm systems of other military bases in Minnesota; and the alarm that was set off was not simply one that noted a breach of the perimeter of the grounds but it also signaled nuclear attack!

Fortunately, it was only at this one base - Volk - that the flight crews were summoned to man their craft. But this was a nuclear equipped squadron and its mission was to intercept incoming Soviet planes flying in over the North Pole.

If they had got off the ground, world war III most likely would have resulted. Fortunately, the command pilot - 27 year-old Lieutenant

Dan Barry - gazed out his cockpit window just in time to note that a light army truck was madly racing toward him down the runway with all of his hazard lights flashing. Could this be a Soviet trick? An airman gone crazy?

Lieutenant Barry thought not. He decided to see what the driver of the truck wanted in such urgency. It was to tell the Lieutenant that the war alert had been triggered by accident and there wasn't an attack underway - despite what then might have been occurring in Cuba.

If the Lieutenant had chosen to continue the mission and simply flew past the onrushing truck with the blazing lights his squadron of F106-A jets that were armed with live 812 pound Genie Missiles could have on their own flown into Soviet airspace and started World War III.

GREAT BLACKOUT OF 1965

Date of occurrence: **November 9, 1965**

Three years after the Cuban Missile crisis. The world was still on edge. Nuclear war was still on most people's minds, despite the resolution to the Cuban matter.

So, what happened without warning on this dark rush hour on the East Coast of the United States in 1965 could have triggered a nuclear exchange.

This was before cell phones. Before the internet and computers. Before even cable television! There were only 3 national television networks in the United States and they were **ALL** headquartered in New York. Now New York was in blackness. And NO ONE knew why or how. Not even President Lyndon Johnson.

Many people assumed that the Northeast had been struck by an ICBM attack from the Soviet Union. But the longer the silence lasted and the less likely any further follow up attacks occurred the less likely it became that a nuclear war had been started. Unless, of course, the United States had quickly counteracted and had totally destroyed the Soviet's ability to wage war. This also seemed highly unlikely.

There was still a cause for high alarm. Remember those monitors that were set in various locations around major cities that warned of nuclear attack due to unusual light variances? A very unusual form of light variance would be a totally darkened New York City at any time. It was still possible that these sensors could signal an ongoing nuclear attack to NORAD or other agencies. Fortunately, this did not happen.

What also made this East Coast power failure so dangerous was that it was the first one. New York had never before gone completely dark.

This was a shock to the nation and the world. What happens when contact with the greatest city in America is lost?

As the night progressed, communication with New York was gradually re-established and it became clear that the East Coast had not been the victim of a Soviet attack. But, it seems a precedent of sorts had been set: the Great Blackout occurred on **November 9th**, a date which will recur often in these pages.

SUNBURST EARTH

Date of occurrence: **May 23, 1967**

One of the primary threats to the earth from a coronal mass ejection of the sun is the catastrophic effect on the world's power grid. Coronal mass ejections are expulsions of huge amounts of magnetic plasma energy from the outer layer of the sun. Basically, it is a mass of the sun being blasted into space. In this case earthward and with potential dire effects. The strongest variety of these are called X-class expulsions.

Solar observers were projecting on May 18, 1967 that such an ejection of plasma and magnetic energy was going to take place. As predicted, just such a huge X-class expulsion did ensue and struck the earth on May 23, 1967 and would continue washing over the planet for a number of days. But it wasn't certain that there would be a next day, for 2 reasons.

One reason is that the entire planet could've been roasted by this massive X-class ejection. The other reason was that nuclear war could easily have been triggered by it, thus destroying the human race.

This was a little publicized possibility at the time which is why few people have heard of the event. One important meteorological memo on this matter was sent to American military officials well before the event would occur; but no one read it until much later!

If this CME did not roast the surface of the planet there was the extreme likelihood that it would disrupt the world's communications grid, including the military, and seriously compromise one specialized type of monitoring device. Not all monitors that are on the lookout for nuclear detonations are the same.

One type of monitor is particularly susceptible to coronal mass ejections because it is triggered by pulses. A quick flash of light and burst of energy as emitted by a CME would register as a nuclear explosion. It wouldn't be powerful enough to overwhelm the system as a true explosion would but it would be clearly more powerful than any "normal" type of detonation or common meteorological event and as such would register as of thermonuclear origin. These monitors were activated by the coronal mass ejection.

The CME of May 23, 1967 was at that time the 2nd worst ever recorded; only the "Carrington Event" of 1859 was more fierce. The 1967 event tipped the world toward nuclear war. The disruption to military communications and defense systems was considerable. Three of the U.S. Air Force's ballistic missile warning radar systems in the northern hemisphere jammed at the same time. This implied enemy sabotage. And sabotage of this nature in itself was considered an act of war.

The Air Force fleet that was in continuous flight near the Northern Polar Regions was placed on standby. The squadrons that were on the ground were scrambled and prepared for war. The country was seconds away from unleashing a nuclear attack on the Soviet Union.

But 2 things prevented this from happening. The first was the quick action taken by an unnamed NORAD official who for some reason was inspired (by what agency we do not know) to read the meteorological report that had been relayed to his department many days previously. This was the report that warned about the CME and the severe effects it would have on all tracking and communications systems, including early warning radar systems. It isn't known what inspired him to finally read it.

Secondly, it was noted by the Pentagon that the Soviets were not on a high state of alert. Also, there weren't any reports from NORAD about hostile ICBM launches. It was also determined at that point that the pulse monitors that had sent warning signals had done so in error.

The fleet of B-52's that had been dispatched to attack the Soviet Union were recalled. And the other squadrons that had been preparing to take off were ordered to stand down. The alert was officially canceled.

The CME of May 1967 spanned several days - from the 23rd to the 27th, oddly the same duration of days as the height of the Cuban Missile Crisis. But, since the actual threat of nuclear war had been met and avoided at the outset of the CME the occurrence of this near nuclear holocaust will be confined to 23rd day of the month. This makes it the second such event to be experienced on the **23rd day of a month**. Another coincidence?

PRE-PROGRAMMED NORAD WAR

Date of occurrence: **November 9, 1979.**

This near nuclear holocaust began at NORAD headquarters on the early morning of November 9, 1979. It's about an actual war game that was set into operation here which nearly destroyed the world.

While the background for the 1983 movie *Wargames* is not officially linked to this scenario, the similarity seems too close to be coincidence. The simulated war game program created by a main character in this movie - Dr. Falken - is very similar to the actual program developed by the super genius Hugh Everett III. The purpose of the real program was to warn that any nuclear exchange would result in Mutual Assured Destruction. The only way to win was not to play. Neither the computer program nor Dr. Everett were given any credit!

Also, in the 1983 movie a member of the U.S. Congress was visiting NORAD during the fictional alarm. In real life, in 1979 U.S. Senator Chuck Percy was visiting NORAD when the real event about to be described happened.

The true genius behind the *Wargames* anti-war computer simulation, Hugh Everett III, was also the developer of the so-called Many Worlds Theory or the multiverse concept. According to this theory, for every choice that a person makes in life an alternate reality is created which his alternate self participates in. At the time he espoused this idea even most other quantum physicists couldn't understand its true overall implications. Thus, he was unable to find employment at education institutions, so Dr. Everett accepted a position at the Pentagon where he developed his MAD theory which demonstrated in detail how any nuclear exchange at any level will ultimately result in full scale nuclear war. So, again, the only way to win was not to play.

Unfortunately, Dr. Everett died at a relatively young age and was no longer living in this reality to be able to accept the wide acclaim that his multiverse concept finally received from the scientific community which finally struggled to catch up to him on an intellectual level.

But, as to the current near holocaust scenario: just before noon on **November 9, 1979** (November 9th again, please note) the main screen at NORAD became filled with images of incoming missiles fired toward the West Coast of the United States by Soviet submarines. A couple of minutes later, they were joined by squadrons of Soviet ICBMs launched from the mainland. The United States was under attack and would be struck within 10 minutes. The proof was on the screen on the wall.

Bombers from the Strategic Air Command were dispatched. ICBM silos across the country were activated. All commercial airlines were about to be grounded. Air Force 1 was en route to pick up President Carter so he would be airborne when war erupted.

The Soviet military observers had to have been shocked and bewildered. There hadn't been any heightened tensions; and THEY knew that they hadn't launched any attacks - not even accidentally! But the United States was certainly preparing an attack. So, the Soviets prepared to respond.

Then...NORAD discovered a problem. While the big screen at its headquarters showed an incoming invasion, no one in the field saw it. None of its remote tracking stations and early warning radar facilities (just like in the 1983 movie) were picking up anything out of the ordinary. Not any of them! Something was wrong!

An immediate diagnostic was run on NORAD's internal computer system and a major error was discovered. The system was being driven by an old program that had been activated to simulate the effects of a Soviet first strike. It was never divulged who activated this program or if it was an accident or true sabotage.

When PAVE PAWS - combined satellite, radar warning system - verified that the Soviet attack was non-existent, all actions taken to repulse it were reversed. Fortunately, the Soviet Union did likewise and called off its defensive response.

A person must wonder: is there an alternative world where this erroneous computer program was not discovered and full scale nuclear war was the result. And, again, this took place on **November 9th** in this continuum.

MR. PETROV SAVES THE WORLD

Date of occurrence: **September 26, 1983**

This event took place in Russia just after midnight (once again at the midnight hour) on September 26, 1983 Russian time (September 25 in most of America). Appropriately it spans 2 of the critical dates in nuclear near devastation history.

At the direct center of this event is the then 44 year-old Soviet Lieutenant Colonel Stanislov Petrov. He was in charge of a secret bunker called Serpukhov-15, an installation outside of Moscow where early warning satellites are monitored which give alerts if an attack was being waged by the United States, its allies, or anyone else.

Stanislav had been peacefully lounging back in his comfortable commander's chair. Suddenly the word START flashed in Russian in great red letters on the control panel in front of him. He bolted forward! This one word message signaled the launch of an enemy ICBM from an American base.

Petrov had to decide whether or not this was a genuine, bona-fide missile launch. No one else could do it; he was alone at the control panel. The reading could've been faulty. There wasn't any technical way to know. At least, not within the 15 allotted seconds in which he had to decide.

Then came a second, third, fourth and fifth missile launching. But why? Why would the United States suddenly attack? Could it be in response to the Soviet military's downing of an unarmed Korean passenger airliner only a couple of weeks earlier. On that flight were 269 innocent people who were killed - some of them Americans.

Would the United States wage all out war over that! The Soviets had claimed that the attack on the airliner had been an accident. According to Petrov, these were the primary thoughts going through his head at the time. Did an all out attack by the United States over an accidental tragedy make sense?

Also, Petrov realized that if a country was going to start a nuclear war it wouldn't do so with a mere 5 missiles. This would assure the attacker's own annihilation. Any true nuclear assault would start with an overwhelming salvo of ICBMs.

Petrov decided that the incoming images and the message of alert that had ruined his peaceful evening were caused by a malfunction in the Soviet tracking system.

Still, he was sitting in the middle of a room of flashing warning lights and blaring sirens. It was Petrov's decision alone on how to officially treat them. He had one hand operating the control system and another hand on the intercom to his superior. Did he report to his awaiting commander that nuclear war had begun, or that this was all caused by a malfunction in their system?

Ultimately, Petrov told the waiting commander that the alert was caused by a malfunctioning system. After the affair was over, Petrov was intensely interrogated at headquarters. He didn't write down anything about the event as it happened. When his superior demanded why not, Petrov simply answered, "Because I don't have 3 hands."

ABLE ARCHER

Date of occurrence: **November 9, 1983**

Able Archer was the code name given to a highly realistic NATO war game exercise held in Europe, spanning the dates of November 7 - 11, 1983. November 9th is noted specifically because it is the moment of highest nuclear peril that occurred during this long event and what happened before the 9th was routine and what happened after that date was also of only common importance. The crisis occurred solely on **November 9, 1983.**

It is critical to point out that this is the 3rd event of this type which occurred on the 9th day of the month of November, though in different years. One has to wonder: what is the probability that 3 near nuclear holocaust scenarios would take place on the same day (9th) of the same month (November), but different years and in different scenarios, rather than during recurring planned exercise programs? How would Dr. Everett have calculated the probabilities?

Able Archer was basically a NATO tactical war game in Europe that was so realistic and expertly planned that it caused the Soviet's and the Warsaw Pact countries to believe that world war III had in actual fact begun. And the Soviets planned to respond!

It's still surprising that the Soviets had been so totally fooled into thinking the event was real. War exercises were common activities in Europe on both sides during the Cold War. But this one by NATO was held at a technically bad time. Just then the Soviets were initiating a new program to spot early indications of a nuclear threat on its European borders and NATO's Able Archer exercise was choreographed to precisely mimic the exact stages that would be used during preparation for a ground based nuclear first strike.

The tactics were acted out on the field in full view of the Soviets whom it was understood would be keeping close watch on the proceedings. Those undertaking the exercise assumed that the tactics were so predictable that the Soviets would see them for what they were - a simple preparation for potential future events. But the Soviet observers believed just the opposite; that the routines being used were so precise and expected that it was a deceptive camouflage for a real attack.

Another item greatly added to the Soviet's suspicion. At this time there was a higher than usual number of ciphered messages passing back and forth among the allies. Not only that, NATO was using a newly developed code that alarmed the Soviets. This was surely intended to cover the actual meaning of the ongoing war exercises, they decided.

There was another matter that was troubling to the Soviet observers. Something else was very different about this NATO war exercise. It was being directly overseen by the top level political leaders of the 3 allied countries. This was exceptionally unusual; in fact, it may never have happened before to anyone;s knowledge.

Able Archer began just as a genuine attack would. The opening stage was unleashing chemical weapons (in simulation) with the purpose of dissolving the front line enemy field forces. This was about a 36 hour process.

Then, in this case on **November 9th** the first use of tactical weapons would begin in order to destroy any defensive positions. This would then lead to a wider scale assault on the Warsaw Pact countries.

What made these events even more threatening to the watching Soviets was that during these procedures NATO Command simulated the normal change of Defcon status, passing from level 5 (peace) to Defcon 1 (war).

In response, the Baltic Military district, which was the initial target zone, placed all of their nuclear forces on the highest alert. All missile silos across the world were activated and the entire Soviet nuclear submarine fleet was placed on standby in preparation for war.

These actions were noted by Lt. General Leonard H. Perroots of NATO Command. He suddenly realized that the Soviets mistook the Able Archer exercises as an actual first step to war. The Lt. General did not have long to respond. He couldn't simply call the Soviets and tell them these were just maneuvers. They certainly wouldn't believe that. He had to act in a way that demonstrated that a shooting war wasn't intended.

The General reacted by not reacting. He maintained NATO's status quo military preparedness stance. Why didn't he actively decrease activity or stand down completely? Because he expected that the Soviets would view this as a ploy designed to trick them into standing down, too. It would be a likely strategy to use if NATO did intend a full scale attack.

Doing nothing out of the ordinary was a very risky procedure, if it could be called that, but there wasn't time for anything else. The military exercise proceeded as normal without any surprises. The Soviets realized that if an attack had been planned by NATO their forces would've had to have reacted to the WARSAW forces being placed on a war footing. To do nothing differently, as NATO was doing, would have resulted in NATO's defeat because the Soviets in their new disposition would have annihilated them. The Soviets were convinced that an attack was not imminent and if one had been planned it had been aborted.

Ultimately, NATO officials contacted Warsaw Pact officials to assure them that ABLE ARCHER had always only been a military exercise. Whether or not they believed this, they did recall their own forces.

Thus, it was a command decision made by Lt. General Perroots not to do anything that prevented the world from engaging in nuclear war again on **November 9** this time in **1983**!

BLACK BRANT XII

Date of occurrence: **October 25, 1995**

Once again, the date of this near nuclear holocaust event occurred on the **25th day** of a month. Just as the third day of the extended Cuban Missile Crisis was the **25th** of the month and the black bear episode in Duluth was on **October 25**, 1962. The Petrov affair might also be termed a 25th day of the month disaster because in America it was the **25th** of September during that event although it had just turned the 26th in the Soviet Union.

The current near holocaust being considered took place on **January 25, 1995**. In order to study the Aurora Borealis over Svalbard, Norway - located halfway between Norway and the North Pole - a group of scientists caused to be launched an experimental 4 stage Black Brant XII rocket toward the Arctic. Because of its very steep trajectory this rocket followed a path that took it across the Minuteman-III nuclear missile silos from North Dakota and over Moscow, the capital of the Soviet Union. This made it appear that the missile had indeed been launched from one of the Minuteman-111 silos in North Dakota.

And, in addition, the Black Brant XII reached an extremely high altitude of 903 miles, which was similar to that of a trident missile launch from a U.S. submarine. The Soviets saw this and feared that a nuclear attack that descended from such a high level, no matter where it originated, would take out their entire radar defense system. Was just such an attack underway?

The Soviet Union went on high alert. President Boris Yeltsin was contacted. He was currently away from the capital but had in his possession the "chegnet" or nuclear briefcase, with which he could launch a nuclear assault from wherever he was. This was its inaugural "use."

But something did not make sense. Like in the Petrov situation, the commander in charge of first strike detection operations was curious as to why only one missile had been launched toward the Soviet Union. This was hardly an overwhelming strike.

And, like another previous situation when an advanced warning had been given but was ignored, this situation was defused when a Soviet official decided to check his official notices of upcoming international space launches. He discovered that a warning was given months ago about this launch, stating that it might be misinterpreted as a nuclear strike for just the reasons it was being misidentified. Fortunately, this information was relayed to President Yeltsin just before he unleashed a nuclear response, which was literally at his fingertips.

It might be recalled that a similar event took place on **May 23, 1967** when a coronal mass ejection almost caused a nuclear holocaust but was defused when an official at NORAD discovered a weeks old document which warned his department of the severe effects that would be caused by the solar emission. Among these severe effects was a malfunction being produced in the early warning radar system, exactly what was then the near cause of war. Nuclear annihilation again was then avoided by the mere reading and forwarding of a document to the proper officials.

Now, as to the date of the **23rd**! The earlier near holocaust just alluded to of **May 23, 1967** was caused by a CME, coronal mass ejection. One of the days of the Cuban missile crisis was on **the 23rd**. And the next event to be examined was also caused by a coronal mass ejection and it happened on the 23rd of July, 2012. Three dates of the **23rd** of a month on which the planet faced near extinction - out of a possible total number of 365 other dates!!!

SOLAR MASS DESTRUCTION

Date of occurrence: **July 23, 2012**

According to the ancient Mayan calendar the world was to have ended on December 21, 2012. If the huge coronal display that occurred in 2012 had been as destructive as it could've been, their calendar would have been early in its prognostication by only 5 months. The world would have ended in July 2012 instead of December. But neither was the prediction correct nor the CME as intense. Both were ***uncomfortably*** close.

Why the world was not destroyed by natural causes on July 23, 2012 is a major cosmic mystery in itself and the primary feature of this section and reason for its inclusion. Why wasn't the earth roasted to a cinder?

This CME was projected to be the most powerful ever hurled at the planet from the sun. It was supposed to consist of a series of blasts, each predecessor clearing the path for the one behind it. They were expected to strike the earth in a series of at least 3 massive shocks, possibly destroying every living thing on the surface of the earth.

Not everyone was warned of its coming. Only a select few were aware of this CME's immense power. The "average" citizen was given scant information about it for fear of wide scale panic. There would be no place for them to hide. There just wouldn't.

Only the elite (super wealthy) and powerful (politicians) were given the facts about the impending doomsday so they could flee to safety. Government officials and their families had shelters prepared for them. The wealthy had long ago constructed their places of subterranean refuge. Since they had places to hide; they were told about the approaching solar tsunami.

But news agencies became suspicious about the evacuation of these "important and influential" people to underground retreats. The government responded by assuring reporters that these were only "open houses" to show off to the top echelon of society how safe our country's powerful people would be during a nuclear attack.

But, assuming this disaster had occurred and the earth was scorched and its population destroyed, who would these special people who had been sheltered and saved govern? The citizenry would be dead. No one seemed to consider that.

But this monster CME did not trigger any ballistic missile warning systems. No early detection radars were triggered. None of the nuclear sensors stationed around major cities were alerted.

This greatest of all CMEs never even reached the earth. And that is the greatest mystery of all. It was heading directly toward the planet, but never made it here. Why not?

No one is really sure. The best explanation given is that it missed the earth by a week. An explanation which makes no sense. The time frame didn't matter. Three massive ejections were blasted outward toward earth and were coming our way. But they never made it; and NOT because they were a week early.

But...suppose this cataclysmic event appeared a week earlier than predicted. So? Its immense size and awesome power would still have deluged the earth. That's how big this CME was.

This CME was so fierce that even the earth's magnetic field could not have deflected it. It would have been overwhelmed.

There **IS NO REASONABLE EXPLANATION** for why earth was spared!! That's why a cover up was attempted by various sources, a cover up that was only needed because news agencies became suspicious. The general populace had no idea about any of it.

Intelligent people at the news agencies knew that the idea that the earth was saved because the CME came a week early is patently absurd. The idea's progenitor is relying on the general public's lack of knowledge of cosmic physics to accept this pitiable explanation. But some real astronomers were connected with the dispensers of news, and knew better.

Another fraudulent scenario is that somehow because one of NASA's early CME detection units in space intercepted the CME this somehow protected the earth from its onslaught. This is too ridiculous of an idea to even consider. How could a tiny, mechanical device provide more of a protection than earth's own magnetic field!! Nonetheless, this was propounded by NASA as an explanation.

The monster CME did not dissipate. It didn't just vanish. So, what happened to it? A force unknown re-directed it onto a path that took it completely outside of the earth's orbit. Ironically, no one ever sought to examine this situation or find a scientific solution. Instead, the mystery was covered up then forgotten.

It is difficult to deny that someone or something saved this planet from devastation. Could it have been aliens? Angels? Or maybe a future earth civilization which monitored the potential disaster that would occur in its past and provided some way to deflect the lethal CME from its path?

This near miss event is the final example of potential worldwide holocausts that were averted presented in this book. And it directly points toward what has become the main theme of this work: the existence of some unknown process or intelligence that has been protecting humankind from destroying itself and by providing this protection at specific times is in this way issuing a notice or warning that up until now seems to have been ignored. Is it the intention of this "force" to cause someone in authority to look at the dates of these events and note how beyond the realm of rational probability it is that they occurred when they did? Coincidence or accident must be ruled out. What is the real answer?

ANALYTICAL VIEW

The revised edition of this book was published in 2023. The span of time covered during which nuclear or nuclear based holocausts could possibly have occurred is from 1952 (when Russia also developed the hydrogen bomb) until current. At this time, it is 71 years. During this 71 year period there have been 11 instances when nuclear annihilation could have been triggered. And during this period there are only 6 dates of the month when these events have occurred.

It's important to once again point out that the events that were chosen to highlight in this book were randomly selected. It just happened to be a fact of reality that they occurred on recurring dates. Each instance was chosen because it was a period of time during which the entire planet could have been destroyed by nuclear or cosmically induced destruction.

If a reader checks any other list of such events, the list will be very similar if not identical to this one. And the dates will be the same. One or 2 additions or subtractions would not change the outcome by very much - especially since the most major of potential nuclear crises would be retained in anyone's list.

Another, closer look, at the specific dates of near catastrophe is in order. Begin with only the dates of the month on which a near nuclear or cosmic cataclysm occurred. There are a total of 11 such events to be examined.

These are the days of the month during which these events took place: 4th, 9th, 23rd, 25th, 26th, and 27th. This is a total of 6 days out of a potential of 365. Of these 6 days, only 2 were without question one-time occurrences - the 4th and the 27th. Because the Petrov event took place just after midnight in Russia on the 26th of the month it could also be considered as taking place on the 25th in the United States.

The numbers get more interesting when recurring dates occur. Of the 11 dates involved in the near holocaust events, 3 of them took place on the 9th, 4 of them took place on the 23rd and 3 (or 4, depending on the Petrov affair) took place on the 25th.

When determining what might be the most dangerous day for inhabitants of the earth based on the above numbers, the dates of the 9th, 23rd and 25th take precedence. Four of the critical dates of near annihilation occurred on the 23rd of a month: Goldsboro, January 23; Cuban Missile Crisis, October 23; Coronal Mass Ejection of May 23, 1967; and the Coronal Mass Ejection of July 23, 2012. These all occurred in different months and different years.

However, 3 near worldwide holocausts all occurred on the date of November 9th: The Great East Coast Power Failure of 1965; Pre-Programmed Norad War, 1979; and the Able Archer Field Maneuvers, 1983. Although they took place in different years, they all happened on November 9th. Hence the title of this book.

The best probability application to use to explain the recurring near nuclear holocausts is the *frequentist* approach. What this theory says is: the relative frequency of an event taking place, observed by the number of repetitions of the target subject event (near nuclear war), is THE measure of the *probability* of that event.

Or, this may be easier to understand: the probability P of an uncertain event A, is defined by the frequency of that event based on previous observations.

Thus, the probability that nuclear war will begin on either the 9th, 23rd, or 25th of the month are the highest based on past observation of occurrence of near nuclear disaster. And, among all of these days, only the 9th of November is a day that specifically repeated 3 times: November 9, 1965; November 9, 1979; and November 9, 1983. It seems that this would make November 9th the most dangerous day out of the year for the world.

POSSIBLE EXPLANATIONS

There is one possible rational explanation for the unusual recurrence of dates as described in this work, but it has only extremely limited application. Some of these near nuclear holocausts have occurred during routinely held military exercises that are performed either in the field or in the cyber world. The exercises held in the field sometimes occurred at regular seasonal periods. But none of them were planned to take place regularly on any date on which a near nuclear holocaust occurred.

The Able Archer field maneuvers that nearly caused a nuclear war were held on November 9, 1983 and coincide datewise with the pre-programmed Norad War of November 9, 1979. But neither of these events - the field maneuvers or the running of the simulation wargame computer program - were performed as regular occurrences that happened on November 9th.

Able Archer took place at various times during the year and the Norad War Game was an accident that should not have been operating at all. If anything, the fact that both near nuclear events happening on the same day of the same month (different years) better served to point toward the unusual aspect of their time of occurrence.

And what of the 3rd event that happened on November 9 (1965) that could've triggered nuclear war, the Great East Coast Power Failure? This bears no relation at all to the other 2 events. In fact, all 3 near nuclear holocausts that took place on a November 9th were random events. As were the 4 that took place on the 23rd day of 4 different months.

When recurring dates are observed to occur this draws attention to them. Someone or some force seems to be attempting to draw attention to these dates of near nuclear holocaust. The reason for such behavior usually is to present a warning. A warning of what may be in the future if nuclear proliferation is allowed to continue and eradication is not ultimately performed. Apparently whoever or whatever the sender of this message is does not possess the physical ability to intervene and must rely on the non-physical means of scenario manipulation - near nuclear holocausts - to direct conscious behavior.

The potential logical sources have already been given.

RECURRENCE

There is at least 1 natural explanation for the recurrence of the near nuclear holocaust events. It was mentioned briefly at the beginning of the book and will be expanded upon now. Yes, all of these events could have happened in the natural course of events without any intervention from any source. They occurred through the process of **apocatastasis**. There are various descriptions of this concept, but the one being used in this book is: an event of history that occurs in a cyclical pattern which is determined by the passage of earth through a certain area of space that had a similar past event take place there.

Our entire solar system is rushing through space. We seem to be heading toward the galaxy of Andromeda. And in this trek through the cosmos our solar system passes through areas of space that it had traveled through in the past. One theory states that the earth performs one orbit of an area of the cosmos every 300,000 years.

At these past crossings - which occurred in the past history of this planet - historical events took place. The theory is that when the earth passes through this location in space, the same type of history will be replayed as that which was originally experienced in that portion of space at that time. Some forms of astrology are based on this concept.

Thus, while passing through an area in space on April 15, 1865 President Lincoln was assassinated. This tragedy impressed a "memory" of this occurrence on that area of space so that the next time earth passed through that zone a terrible murder would take place, too. It wouldn't necessarily be a presidential assassination, but it would be something of a similar, striking striking nature. Maybe the killing of Archduke Franz Ferdinand which started World War One.

How does this relate to the recurrence of near nuclear holocausts? In reality, not well; but, it may still do so. It relates in the sense that at one time in the past 300,000 years earth may have experienced a period of nuclear peril which the planet has been re-experiencing over the last 71 years. Maybe a past, distant civilization once destroyed itself through nuclear war while passing through this same zone in space through which our planet is now passing. Apocatastasis might be made to fit.

However, for now, there seems to be a scant rational explanation for the recurrence of the episodes of near nuclear devastation. It simply appears that someone or something is trying to save humanity from itself by exposing the type of doom it is facing.

THE END

?

www.ingramcontent.com/pod-product-compliance
Lightning Source LLC
Chambersburg PA
CBHW072157020426
42334CB00018B/2045